TRACING NUMBER

PRESCHOOLERS PRACTICE WRITING WORKBOOK, KIDS AGES 3-5

Book 21

https://tinyurl.com/tracingletterforkidsages3-5

COPYRIGHT NOTICE

Copyright © 2017 by **TRACING NUMBER PRESCHOOLERS PRACTICE WRITING WORKBOOK,KIDS AGES 3-5**.

All rights reserved. This book or any portion thereof may not be reproduced or used in any manner whatsoever without the express written permission of the publisher except for the use of brief quotations in a book review.

CONTENTS

INTRODUCTION	1
HOW TO USE THIS BOOK	2
SMALL NUMBER WORDS EXERCISE 1	*3*
SMALL NUMBER WORDS EXERCISE 2	*4*
SMALL NUMBER WORDS EXERCISE 3	*5*
SMALL NUMBER WORDS EXERCISE 4	*6*
SMALL NUMBER WORDS EXERCISE 5	*7*
SMALL NUMBER WORDS EXERCISE 6	*8*
SMALL NUMBER WORDS EXERCISE 7	*9*
SMALL NUMBER WORDS EXERCISE 8	*10*
SMALL NUMBER WORDS EXERCISE 9	*11*
SMALL NUMBER WORDS EXERCISE 10	*12*
SMALL NUMBER WORDS EXERCISE 11	*13*
SMALL NUMBER WORDS EXERCISE 12	*14*
SMALL NUMBER WORDS EXERCISE 13	*15*
SMALL NUMBER WORDS EXERCISE 14	*16*
SMALL NUMBER WORDS EXERCISE 15	*17*
SMALL NUMBER WORDS EXERCISE 16	*18*
SMALL NUMBER WORDS EXERCISE 17	*19*
SMALL NUMBER WORDS EXERCISE 18	*20*
SMALL NUMBER WORDS EXERCISE 19	*21*
SMALL NUMBER WORDS EXERCISE 20	*22*
SMALL NUMBER WORDS EXERCISE 21	*23*
SMALL NUMBER WORDS EXERCISE 22	*24*
SMALL NUMBER WORDS EXERCISE 23	*25*
SMALL NUMBER WORDS EXERCISE 24	*26*
SMALL NUMBER WORDS EXERCISE 25	*27*
SMALL NUMBER WORDS EXERCISE 26	*28*
SMALL NUMBER WORDS EXERCISE 27	*29*
SMALL NUMBER WORDS EXERCISE 28	*30*
SMALL NUMBER WORDS EXERCISE 29	*31*
SMALL NUMBER WORDS EXERCISE 30	*32*
SMALL NUMBER WORDS EXERCISE 31	*33*
SMALL NUMBER WORDS EXERCISE 32	*34*
SMALL NUMBER WORDS EXERCISE 33	*35*

SMALL NUMBER WORDS EXERCISE 34	*36*
SMALL NUMBER WORDS EXERCISE 35	*37*
SMALL NUMBER WORDS EXERCISE 36	*38*
SMALL NUMBER WORDS EXERCISE 37	*39*
SMALL NUMBER WORDS EXERCISE 38	*40*
SMALL NUMBER WORDS EXERCISE 39	*41*
SMALL NUMBER WORDS EXERCISE 40	*42*
SMALL NUMBER WORDS EXERCISE 41	*43*
SMALL NUMBER WORDS EXERCISE 42	*44*
SMALL NUMBER WORDS EXERCISE 43	*45*
SMALL NUMBER WORDS EXERCISE 44	*46*
SMALL NUMBER WORDS EXERCISE 45	*47*
SMALL NUMBER WORDS EXERCISE 46	*48*
SMALL NUMBER WORDS EXERCISE 47	*49*
SMALL NUMBER WORDS EXERCISE 48	*50*
SMALL NUMBER WORDS EXERCISE 49	*51*
SMALL NUMBER WORDS EXERCISE 50	*52*
SMALL NUMBER WORDS EXERCISE 51	*53*
SMALL NUMBER WORDS EXERCISE 52	*54*
SMALL NUMBER WORDS EXERCISE 53	*55*
SMALL NUMBER WORDS EXERCISE 54	*56*
SMALL NUMBER WORDS EXERCISE 55	*57*
SMALL NUMBER WORDS EXERCISE 56	*58*
SMALL NUMBER WORDS EXERCISE 57	*59*
SMALL NUMBER WORDS EXERCISE 58	*60*
SMALL NUMBER WORDS EXERCISE 59	*61*
SMALL NUMBER WORDS EXERCISE 60	*62*
SMALL NUMBER WORDS EXERCISE 61	*63*
SMALL NUMBER WORDS EXERCISE 62	*64*
SMALL NUMBER WORDS EXERCISE 63	*65*
SMALL NUMBER WORDS EXERCISE 64	*66*
SMALL NUMBER WORDS EXERCISE 65	*67*
SMALL NUMBER WORDS EXERCISE 66	*68*
SMALL NUMBER WORDS EXERCISE 67	*69*
SMALL NUMBER WORDS EXERCISE 68	*70*
SMALL NUMBER WORDS EXERCISE 69	*71*
SMALL NUMBER WORDS EXERCISE 70	*72*
SMALL NUMBER WORDS EXERCISE 71	*73*
SMALL NUMBER WORDS EXERCISE 72	*74*

SMALL NUMBER WORDS EXERCISE 73	*75*
SMALL NUMBER WORDS EXERCISE 74	*76*
SMALL NUMBER WORDS EXERCISE 75	*77*
SMALL NUMBER WORDS EXERCISE 76	*78*
SMALL NUMBER WORDS EXERCISE 77	*79*
SMALL NUMBER WORDS EXERCISE 78	*80*
SMALL NUMBER WORDS EXERCISE 79	*81*
SMALL NUMBER WORDS EXERCISE 80	*82*
CHECK OUT MORE BOOKS BELOW	**85**

INTRODUCTION

A child who learns to trace NUMBERS at home, at the early age with their loving parent or caregiver, grows in self-confidence and independence. This promotes greater maturity, increases discipline and lays the basis for moral literacy. A child who begins with early learning books has a distinct advantage over his or her peers. One of the big advantages being there is no psychological pressure. The reason why parent understand why early learning is important..

HOW TO USE THIS BOOK

Practice, Practice, Practice makes life easier and worthwhile so train your child analytical mind by tracing letters and numbers in the alphabet the conventional way through handwritings as the saying said: "**Young children need writing to help them learn about reading, they need reading to help them learn about writing; and they need oral language to help them learn about both.**"

Make your tracing at your own style and convenience all individuals has its own uniqueness. *

SMALL NUMBER WORDS EXERCISE 1

seventy seven

eighty nine

twenty seven

ninety eight

sixteen

thirty two

SMALL NUMBER WORDS EXERCISE 2

sixty six

thirty seven

seventy one

twenty seven

twenty three

eighty four

SMALL NUMBER WORDS EXERCISE 3

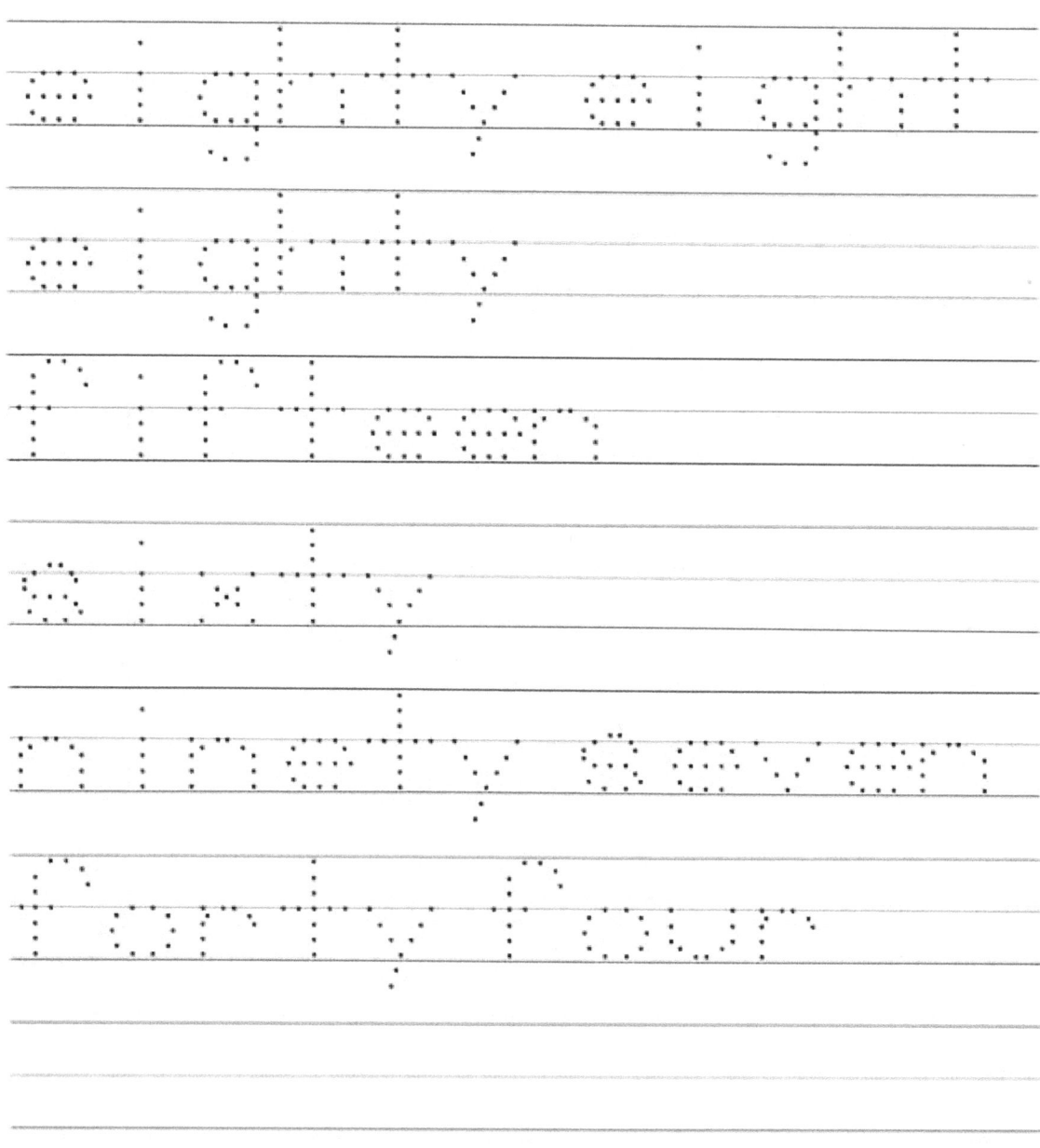

eighty eight

eighty

fifteen

sixty

ninety seven

forty four

SMALL NUMBER WORDS EXERCISE 4

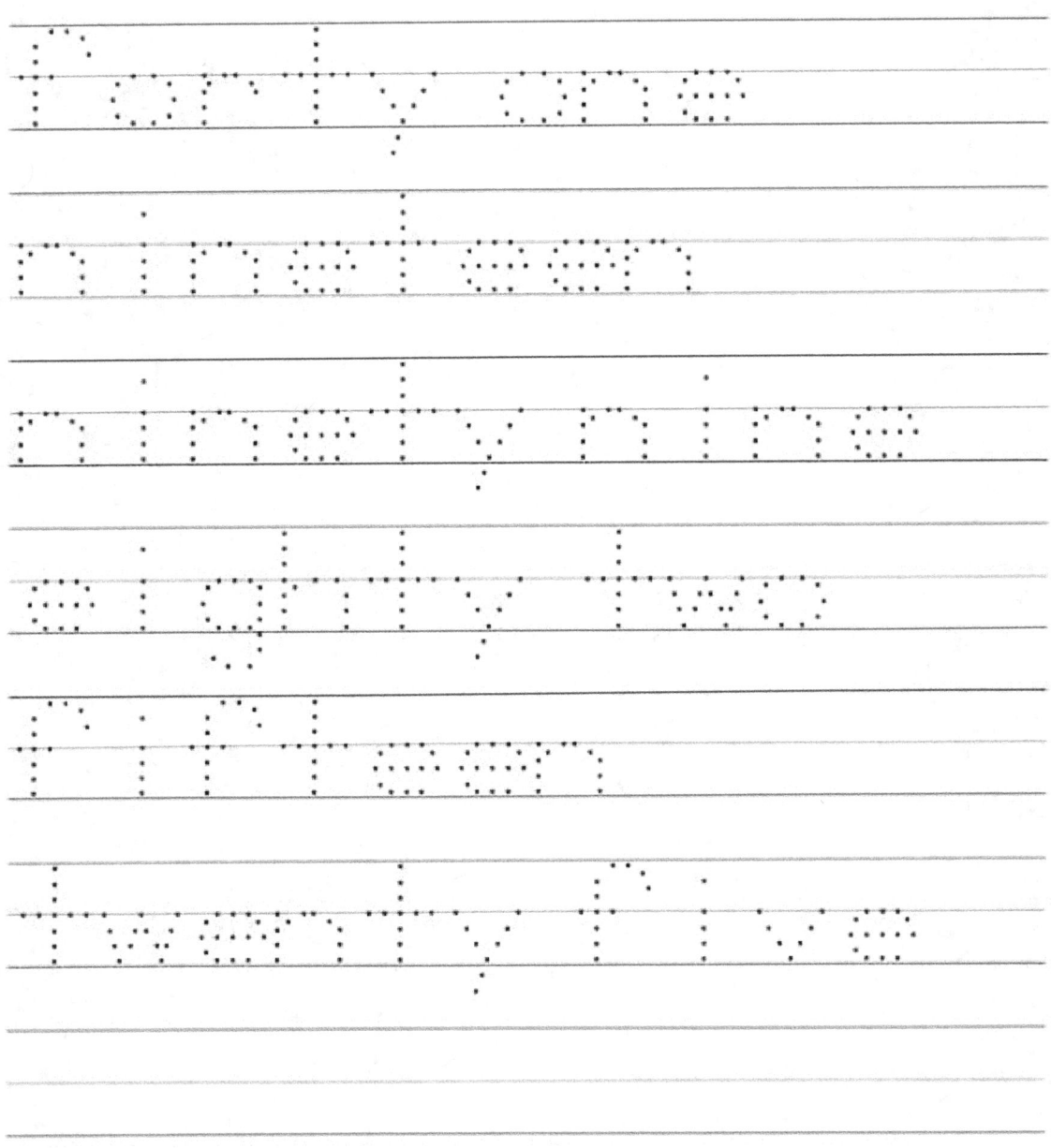

forty one

nineteen

ninety nine

eighty two

fifteen

twenty five

SMALL NUMBER WORDS EXERCISE 5

fifteen

fifty seven

four

seventy one

eighty seven

ninety four

SMALL NUMBER WORDS EXERCISE 6

one hundred

seventy five

eighty four

seventy two

five

forty three

SMALL NUMBER WORDS EXERCISE 7

SMALL NUMBER WORDS EXERCISE 8

three

ninety seven

fifty one

thirty four

seventy five

thirty one

SMALL NUMBER WORDS EXERCISE 9

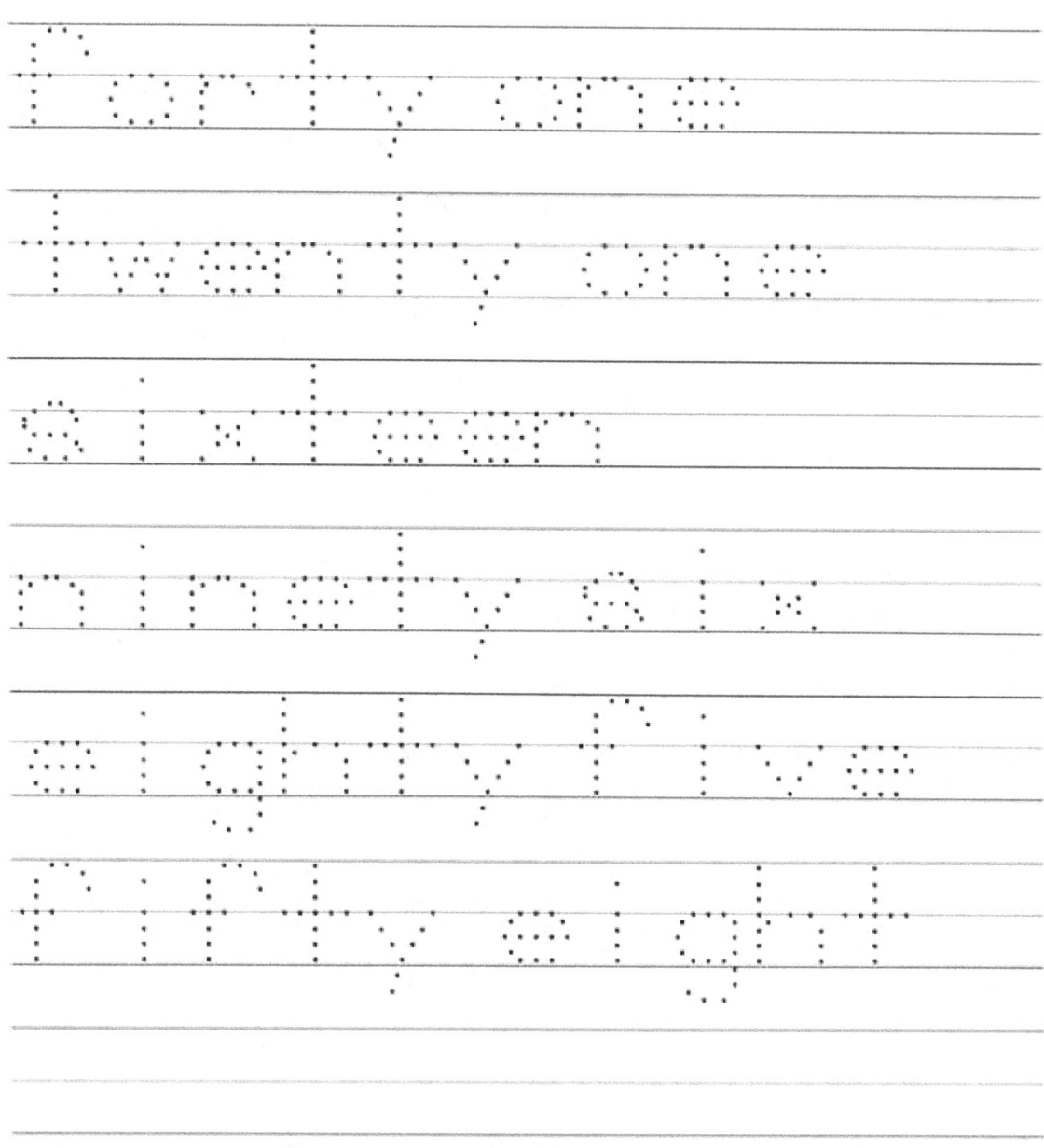

forty one

twenty one

sixteen

ninety six

eighty five

fifty eight

SMALL NUMBER WORDS EXERCISE
10

fifty two

seventy two

sixty eight

eighty eight

sixty nine

two

SMALL NUMBER WORDS EXERCISE
11

fourteen

fifty five

ninety five

eighty seven

sixty

forty nine

SMALL NUMBER WORDS EXERCISE 12

eighty seven

forty two

forty six

seventy

nine

ninety three

SMALL NUMBER WORDS EXERCISE 13

ninety-two

fifty-nine

seventy

eighty-five

fifteen

ninety-three

SMALL NUMBER WORDS EXERCISE 14

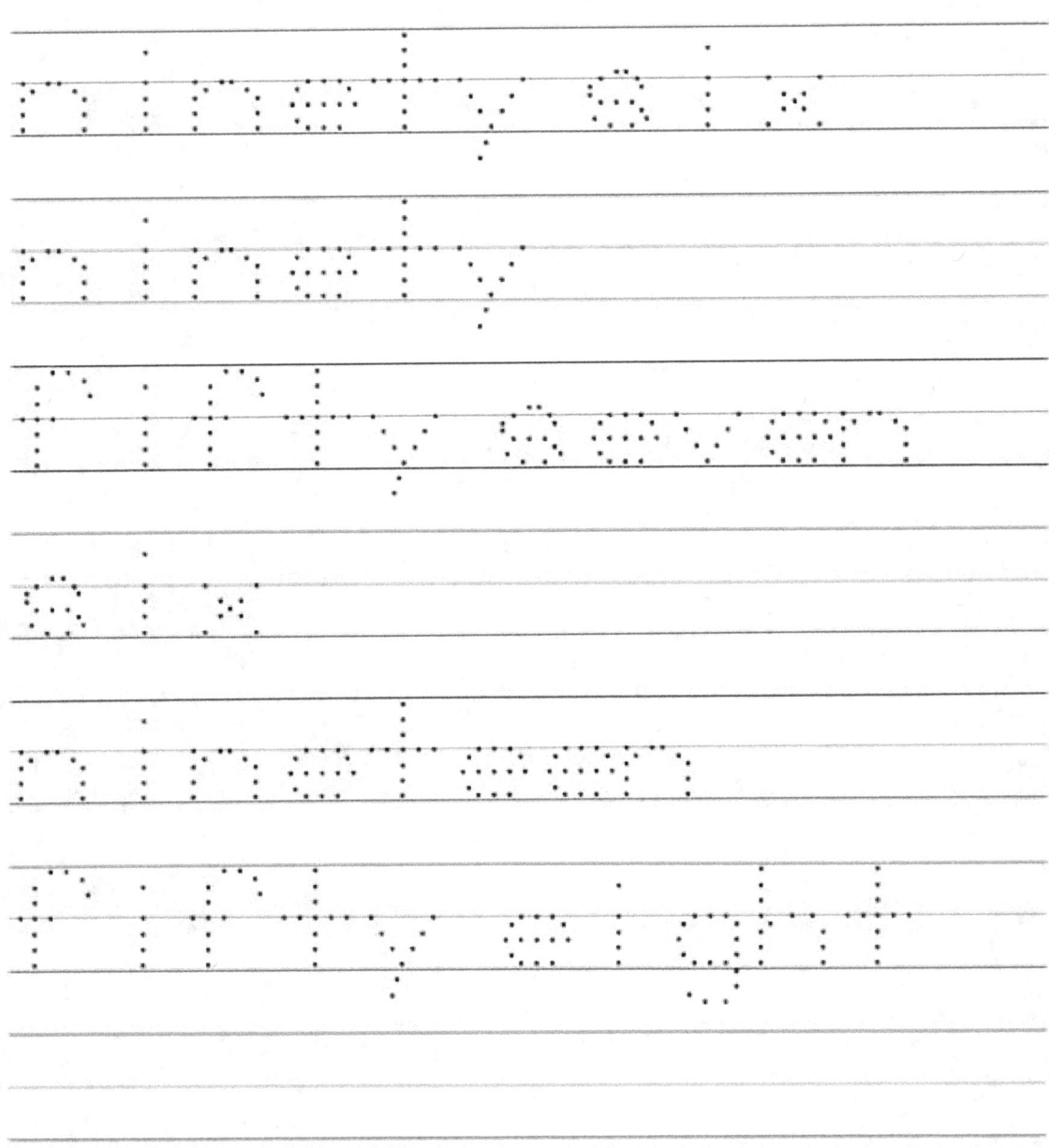

ninety six

ninety

fifty seven

six

nineteen

fifty eight

SMALL NUMBER WORDS EXERCISE 15

twenty

seventy five

ninety two

ninety nine

seventy eight

sixty one

SMALL NUMBER WORDS EXERCISE
16

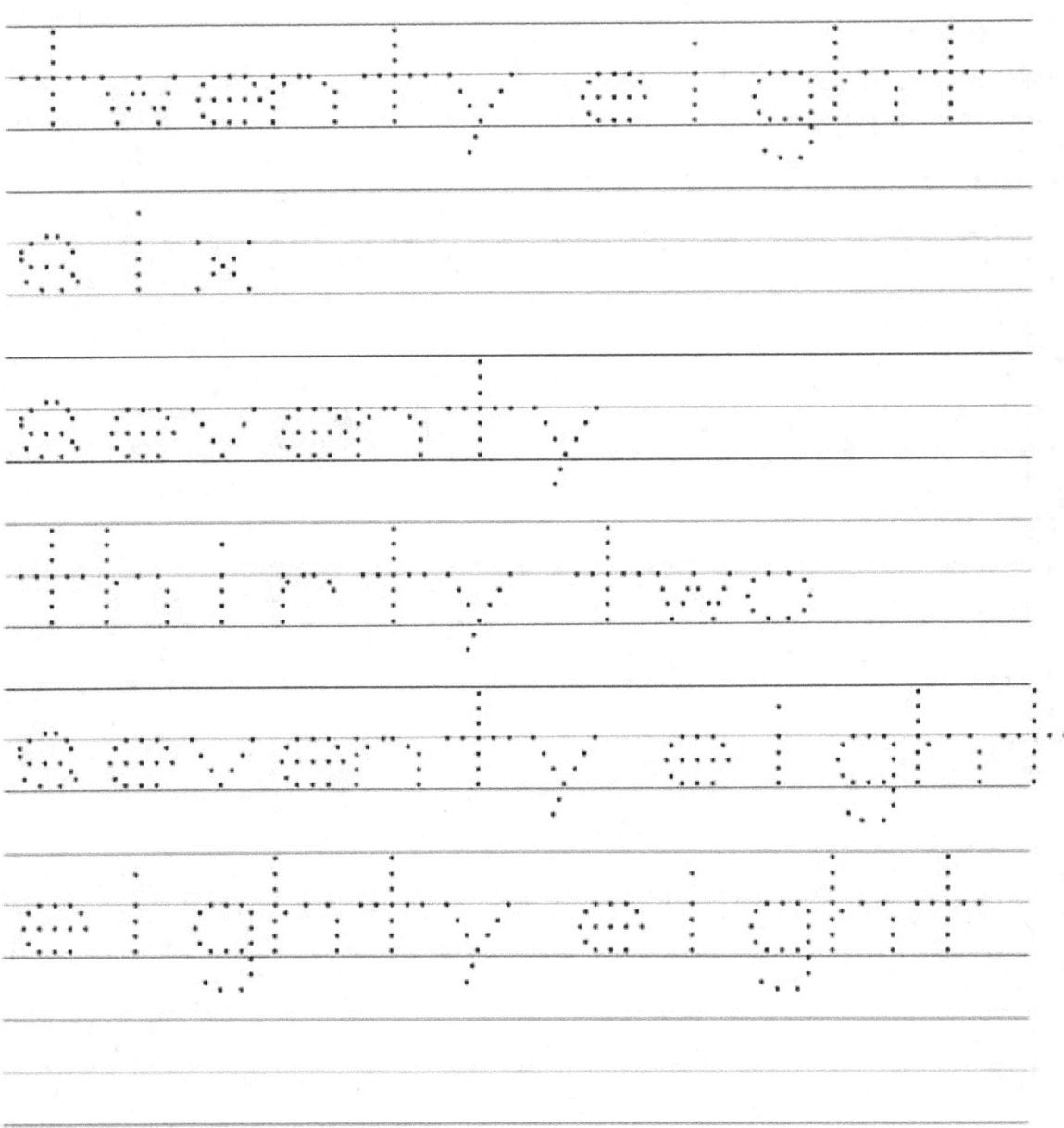

twenty eight

six

seventy

thirty two

seventy eight

eighty eight

SMALL NUMBER WORDS EXERCISE 17

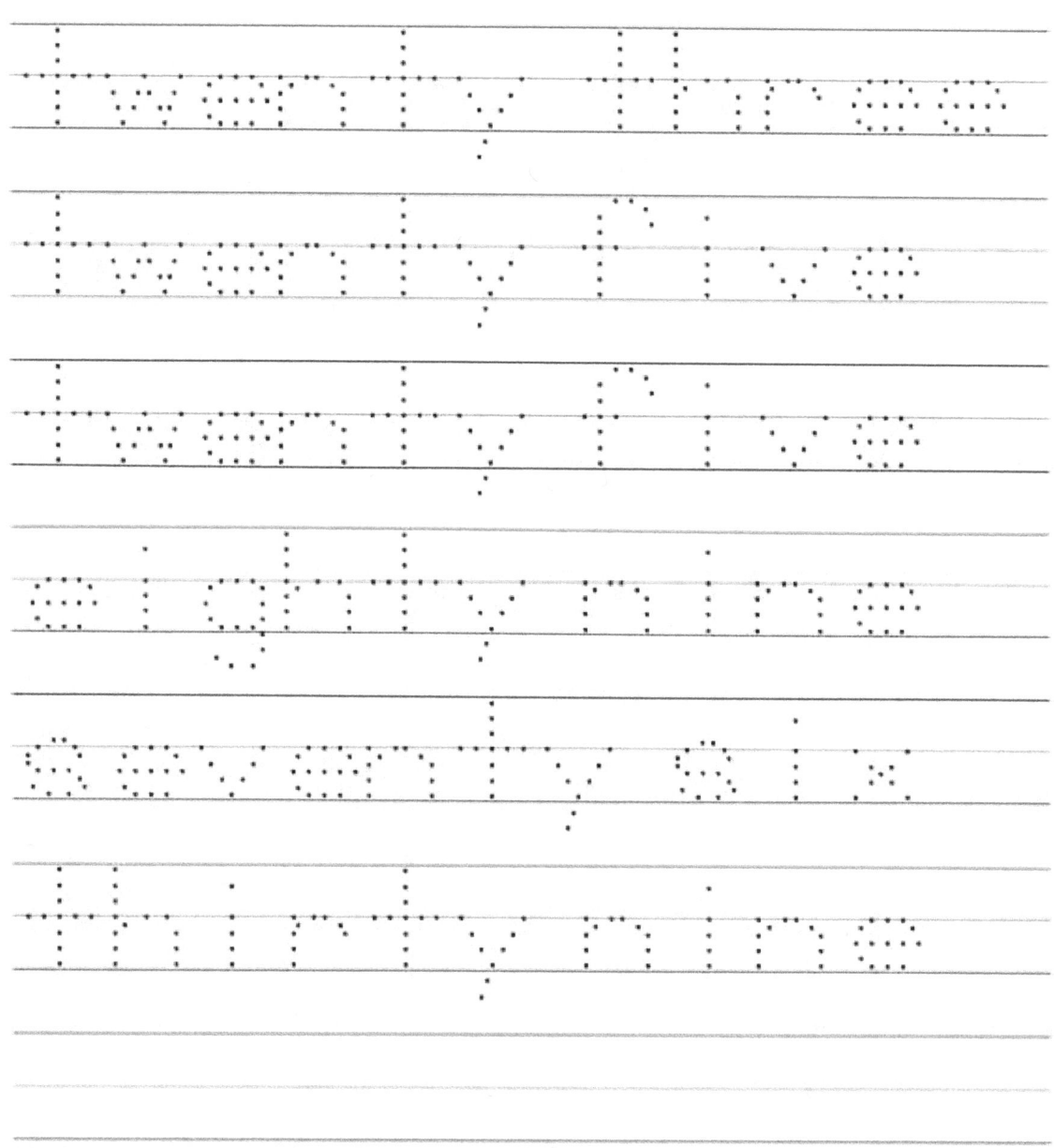

twenty three

twenty five

twenty five

eighty nine

seventy six

thirty nine

SMALL NUMBER WORDS EXERCISE
18

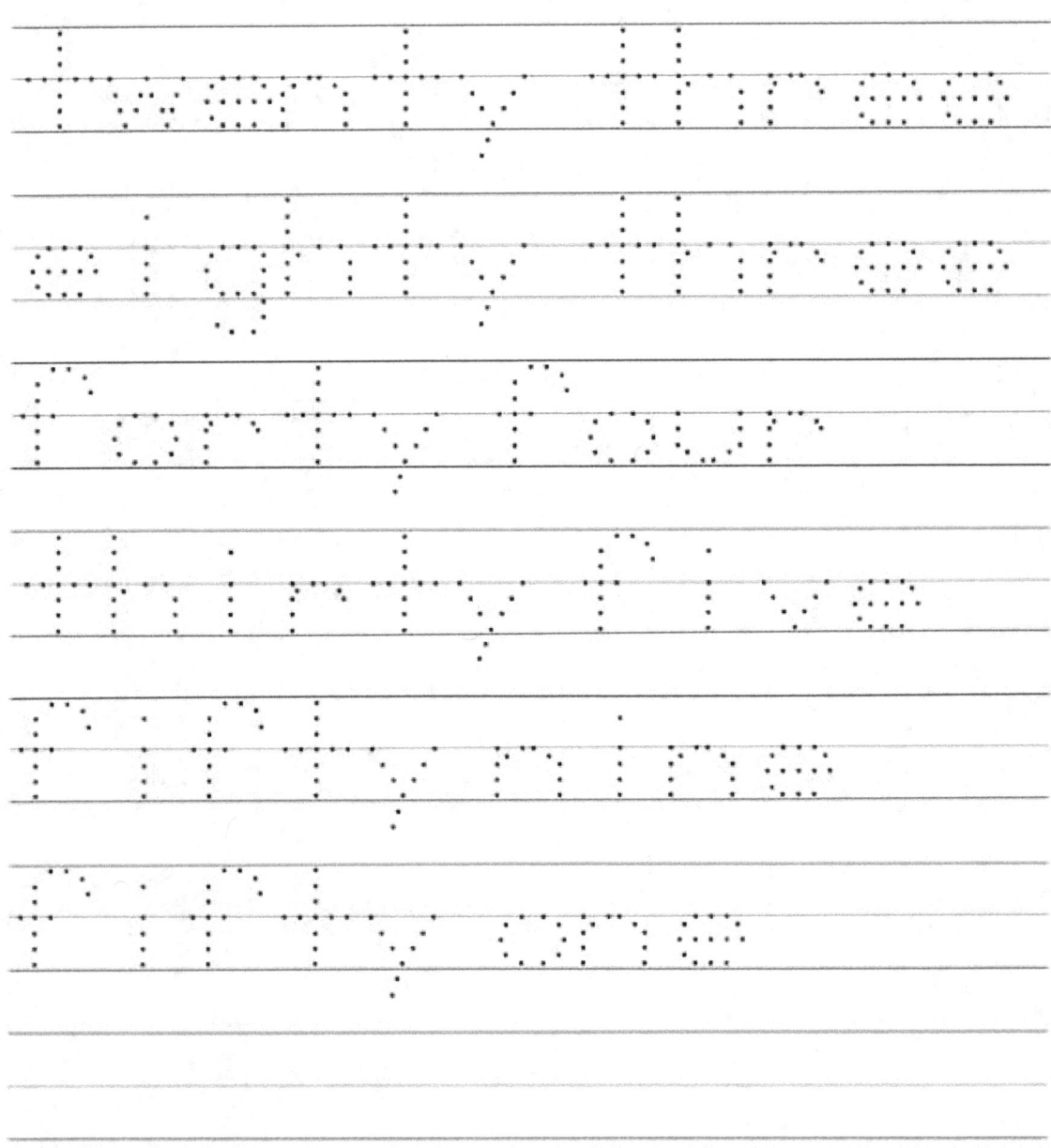

twenty three

eighty three

forty four

thirty five

fifty nine

fifty one

SMALL NUMBER WORDS EXERCISE
19

seventy four

five

ten

sixty one

sixty four

seventy six

SMALL NUMBER WORDS EXERCISE 20

thirteen

fifty eight

forty five

seventy three

sixty two

ninety eight

SMALL NUMBER WORDS EXERCISE
21

fifty three

seven

ninety three

thirty eight

eighty eight

eighty seven

SMALL NUMBER WORDS EXERCISE 22

sixteen

eighty eight

forty six

sixty three

eleven

eighteen

SMALL NUMBER WORDS EXERCISE 23

twenty

eleven

seventy

eleven

five

sixty seven

SMALL NUMBER WORDS EXERCISE 24

fifty one

fifty

twenty

eleven

seventy four

sixty four

SMALL NUMBER WORDS EXERCISE 25

fifty one

fifty

twenty

eleven

seventy four

sixty four

SMALL NUMBER WORDS EXERCISE
26

forty one

fifty four

fifty eight

fifty

seventeen

eighty two

SMALL NUMBER WORDS EXERCISE 27

nineteen

twenty one

twenty nine

eighty two

eighteen

forty five

SMALL NUMBER WORDS EXERCISE
28

twenty eight

twenty nine

one hundred

fourteen

fifty eight

forty six

SMALL NUMBER WORDS EXERCISE 29

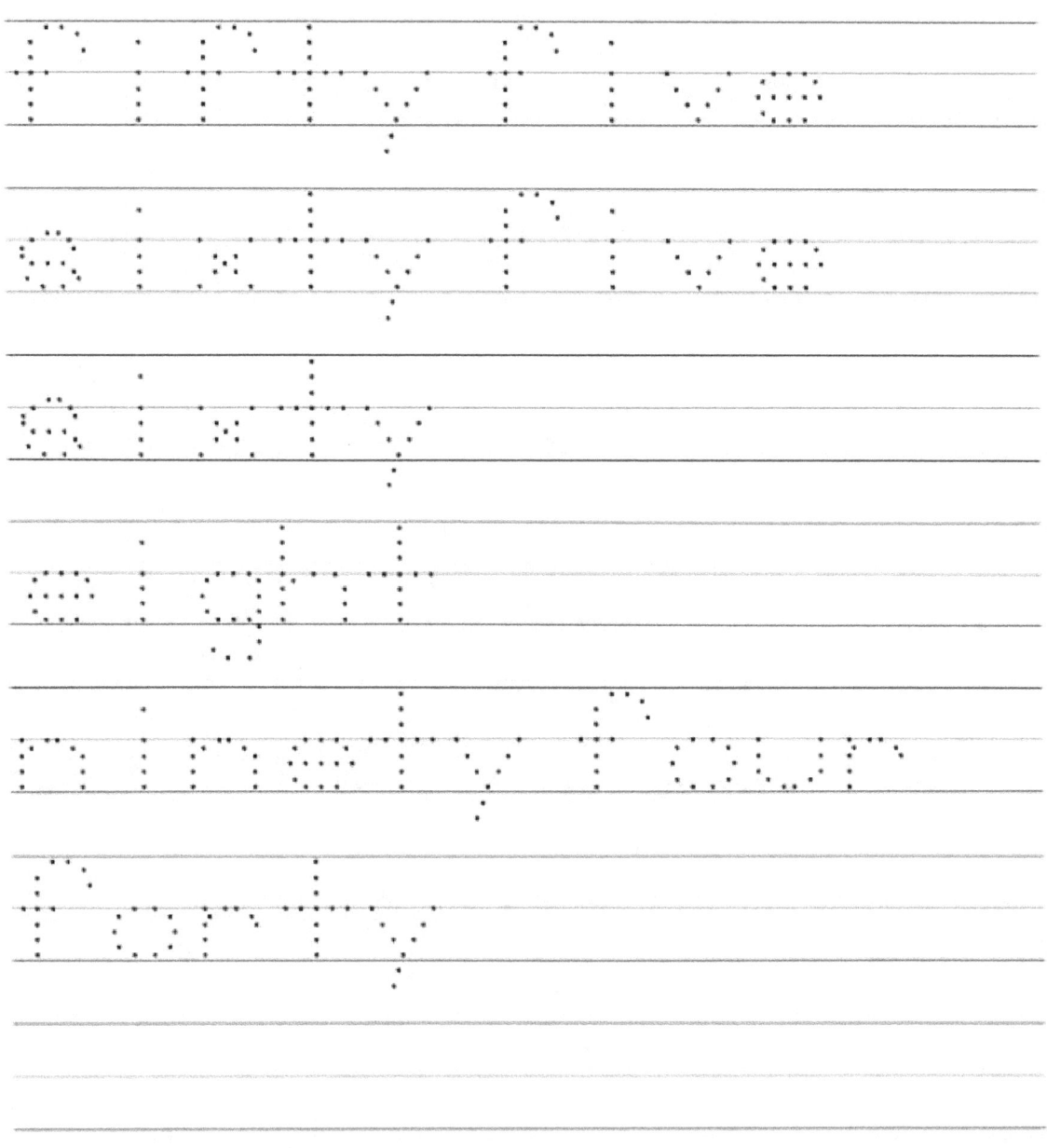

fifty five

sixty five

sixty

eight

ninety four

forty

SMALL NUMBER WORDS EXERCISE
30

seven

forty three

eighty six

thirty eight

two

seventy two

SMALL NUMBER WORDS EXERCISE
31

eighty five

thirteen

thirteen

eight

seventy four

thirty five

SMALL NUMBER WORDS EXERCISE 32

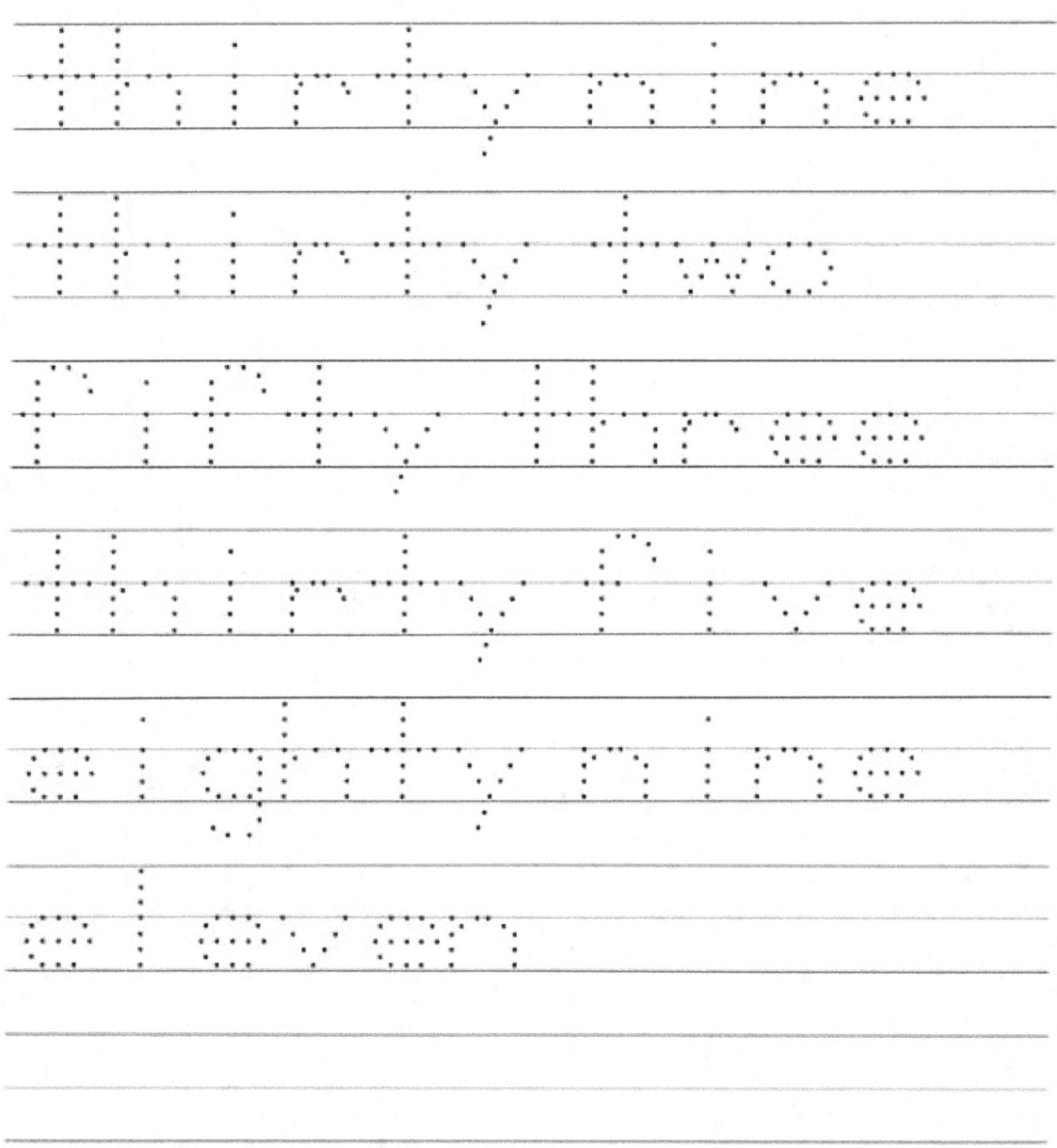

thirty nine
thirty two
fifty three
thirty five
eighty nine
eleven

SMALL NUMBER WORDS EXERCISE 33

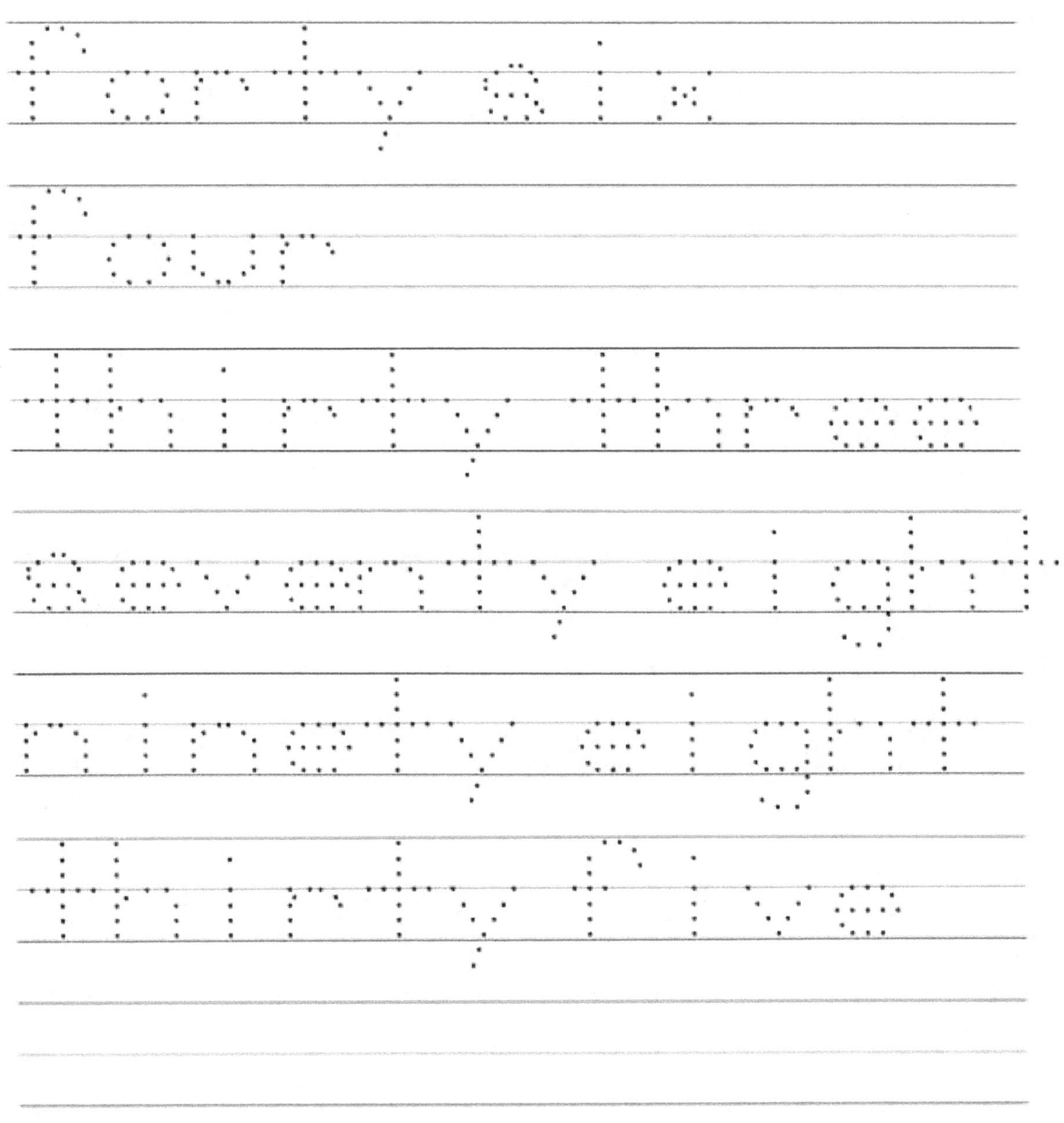

forty six

four

thirty three

seventy eight

ninety eight

thirty five

SMALL NUMBER WORDS EXERCISE 34

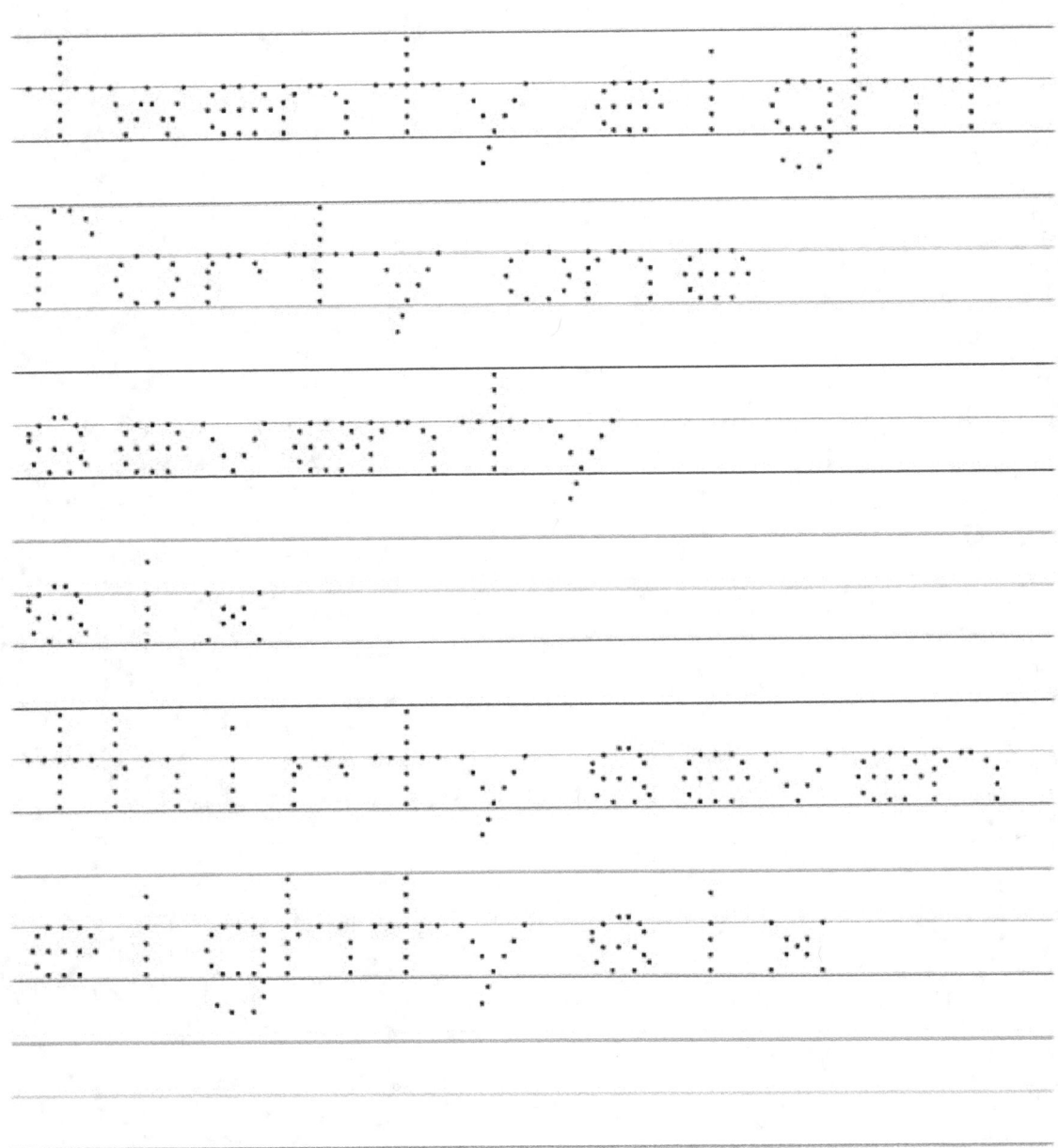

twenty eight

forty one

seventy

six

thirty seven

eighty six

SMALL NUMBER WORDS EXERCISE
35

fifty three

forty three

thirty four

one

fifty one

fifty two

SMALL NUMBER WORDS EXERCISE
36

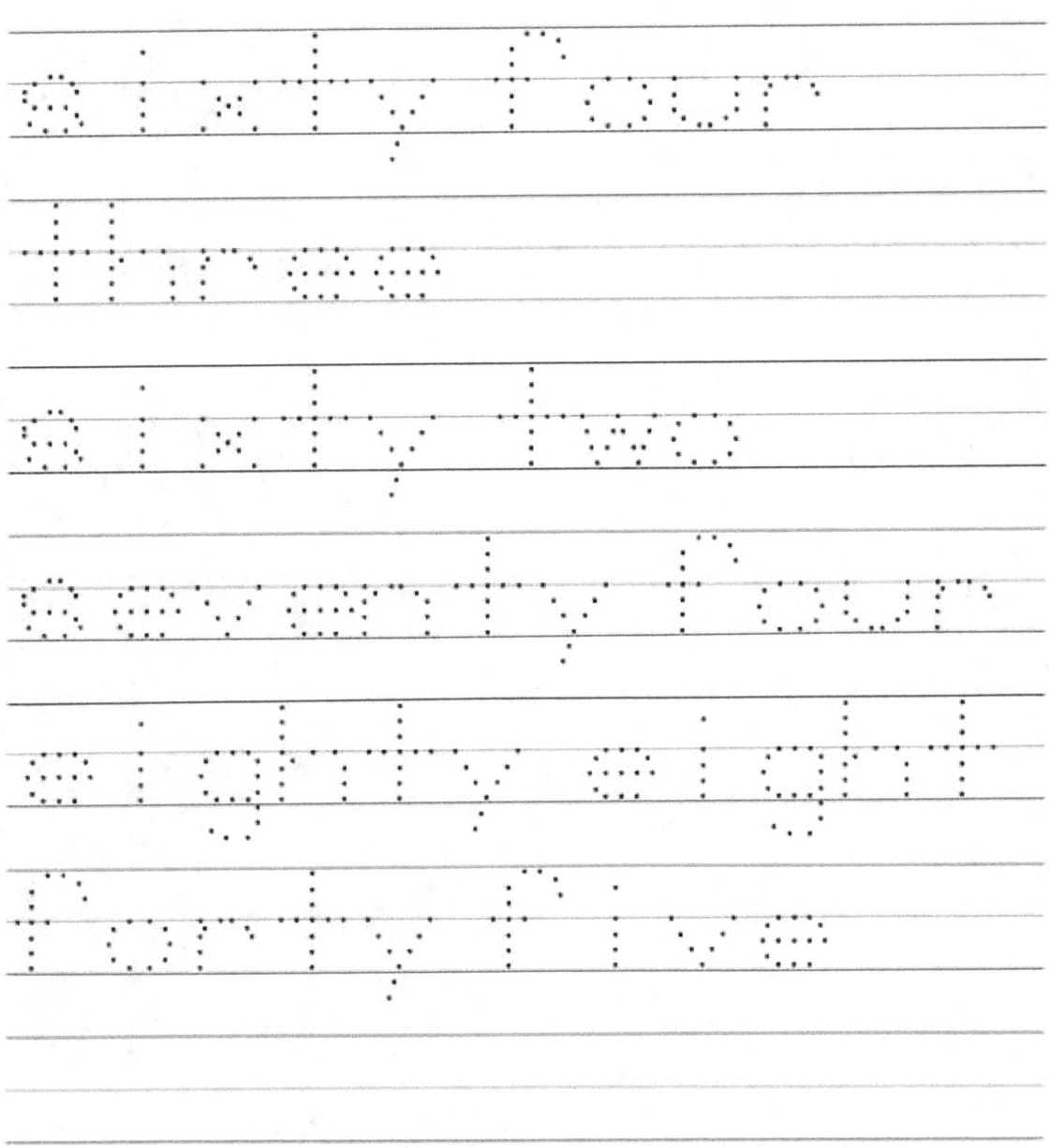

sixty four

three

sixty two

seventy four

eighty eight

forty five

SMALL NUMBER WORDS EXERCISE
37

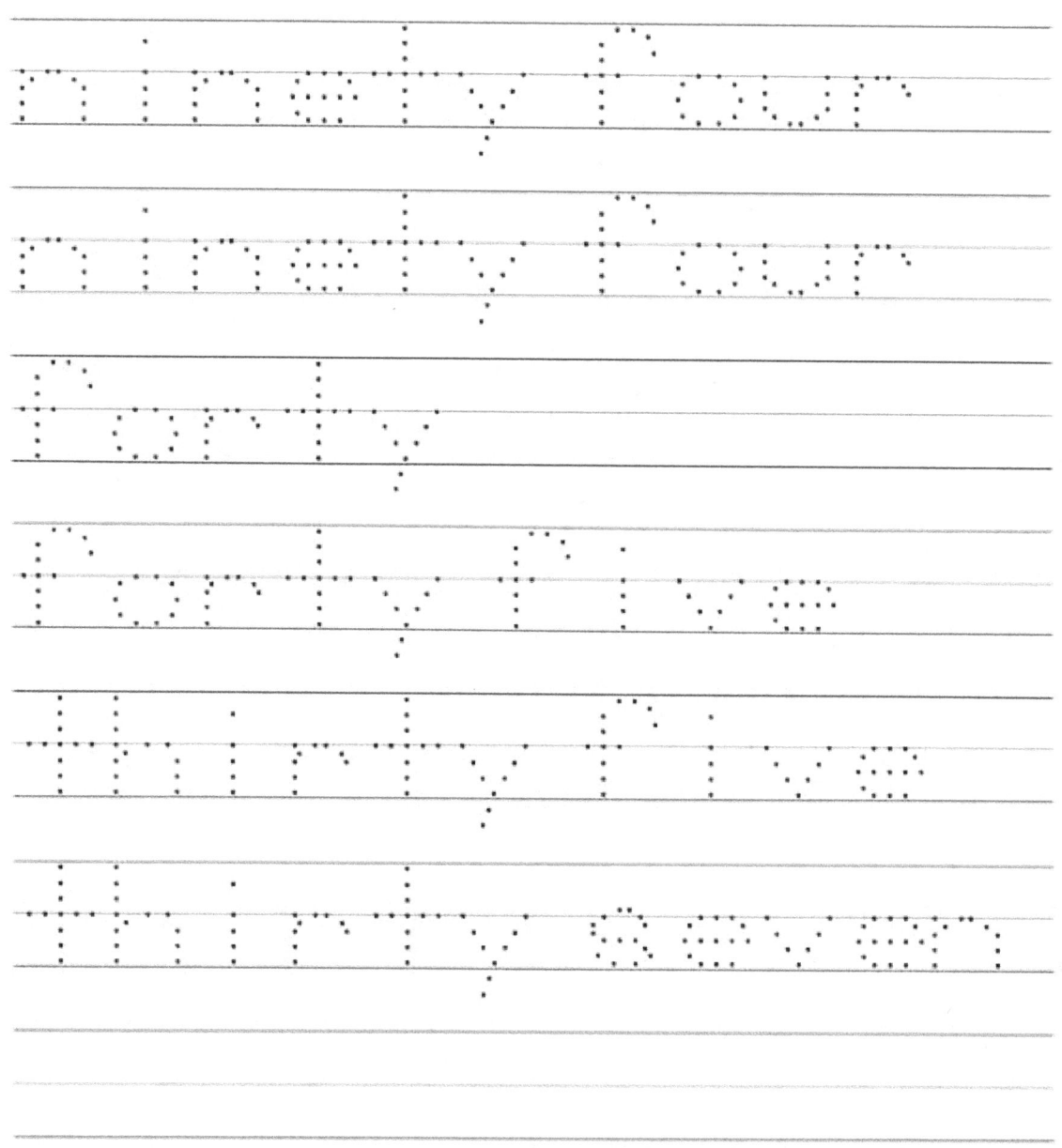

ninety four

ninety four

forty

forty five

thirty five

thirty seven

SMALL NUMBER WORDS EXERCISE 38

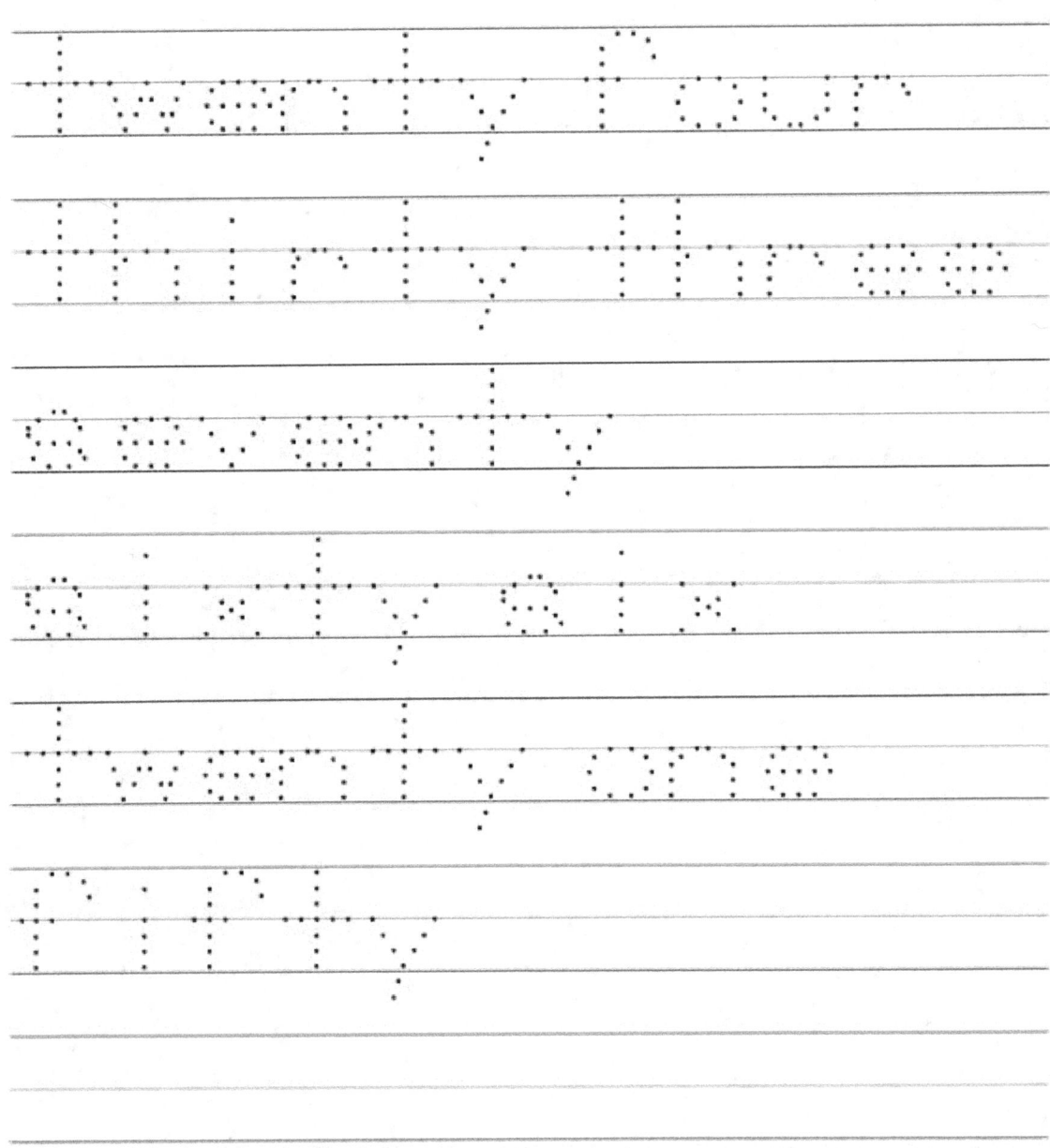

SMALL NUMBER WORDS EXERCISE
39

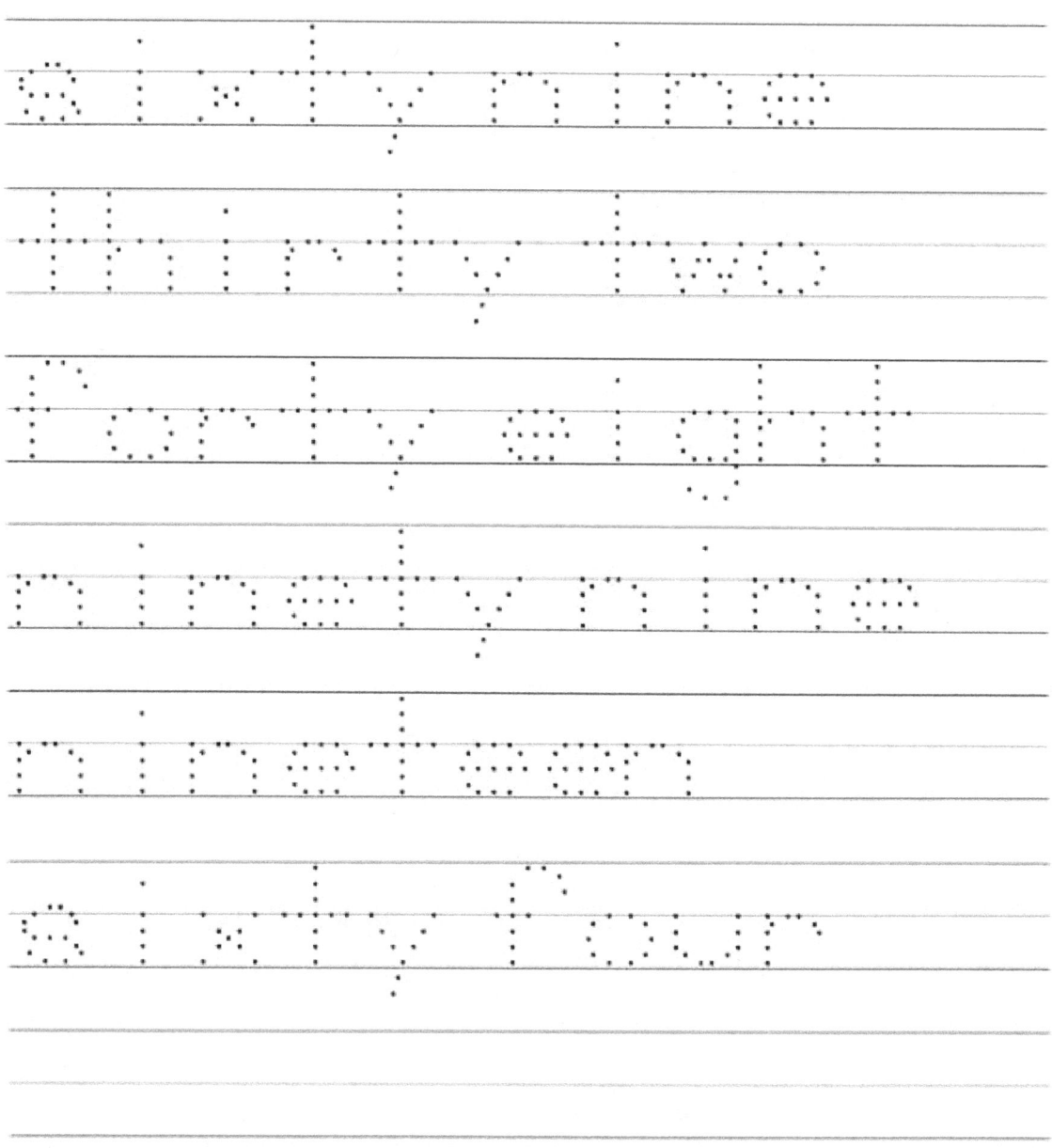

sixty nine

thirty two

forty eight

ninety nine

nineteen

sixty four

SMALL NUMBER WORDS EXERCISE
40

forty two

nineteen

seventeen

twenty three

thirty nine

fifty eight

SMALL NUMBER WORDS EXERCISE
41

twenty five

thirteen

sixty three

twenty one

fifty five

forty nine

SMALL NUMBER WORDS EXERCISE
42

SMALL NUMBER WORDS EXERCISE
43

eighty nine
twenty eight
twenty nine
sixty four
seventy two
ninety five

SMALL NUMBER WORDS EXERCISE
44

seventy five

ninety four

seventy four

twenty four

twenty six

seventy three

SMALL NUMBER WORDS EXERCISE 45

seventeen

twenty nine

thirty

thirty seven

twenty three

ninety eight

SMALL NUMBER WORDS EXERCISE 46

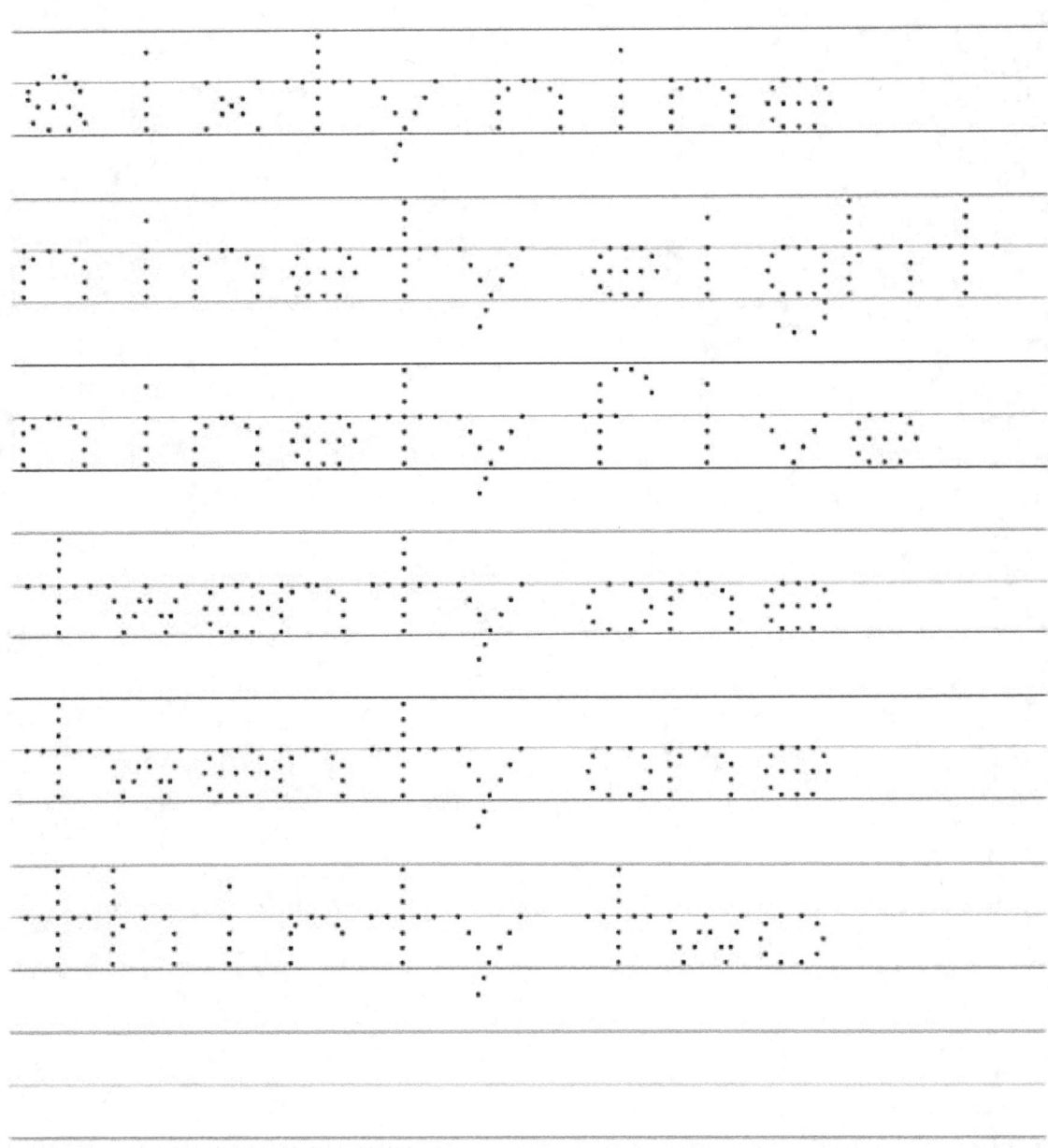

sixty nine

ninety eight

ninety five

twenty one

twenty one

thirty two

SMALL NUMBER WORDS EXERCISE
47

thirty

eighty six

two

eight

twenty eight

sixty two

SMALL NUMBER WORDS EXERCISE 48

seventeen

thirty

five

seventeen

ninety-two

thirty-one

SMALL NUMBER WORDS EXERCISE 49

thirty one

one

twenty nine

fifty two

ninety two

sixty three

SMALL NUMBER WORDS EXERCISE 50

nine

thirty nine

forty six

fifty six

ninety six

fourteen

SMALL NUMBER WORDS EXERCISE
51

seventy three

eighty five

eighty six

forty one

seventy

twenty nine

SMALL NUMBER WORDS EXERCISE
52

two

seventy one

seven

eighty four

fifty nine

fourteen

SMALL NUMBER WORDS EXERCISE 53

eighty one

sixty seven

three

fifty nine

twenty nine

eighty nine

SMALL NUMBER WORDS EXERCISE
54

seventy four

ninety seven

eighteen

seventy seven

forty five

six

SMALL NUMBER WORDS EXERCISE
55

forty three
forty two
five
thirty two
forty six
eight

SMALL NUMBER WORDS EXERCISE
56

three

twenty

twenty three

fifty seven

twenty three

fifty six

SMALL NUMBER WORDS EXERCISE
57

SMALL NUMBER WORDS EXERCISE
58

twenty nine

forty four

one hundred

forty three

sixty three

six

SMALL NUMBER WORDS EXERCISE
59

ninety eight

thirty four

seven

twenty

sixteen

fifty five

SMALL NUMBER WORDS EXERCISE
60

five

eighty six

fifty eight

thirty two

twenty one

forty one

SMALL NUMBER WORDS EXERCISE
61

twenty nine

sixty eight

thirty

forty seven

sixty

sixty one

SMALL NUMBER WORDS EXERCISE
62

twenty two

four

thirty one

seventeen

twenty one

forty three

SMALL NUMBER WORDS EXERCISE
63

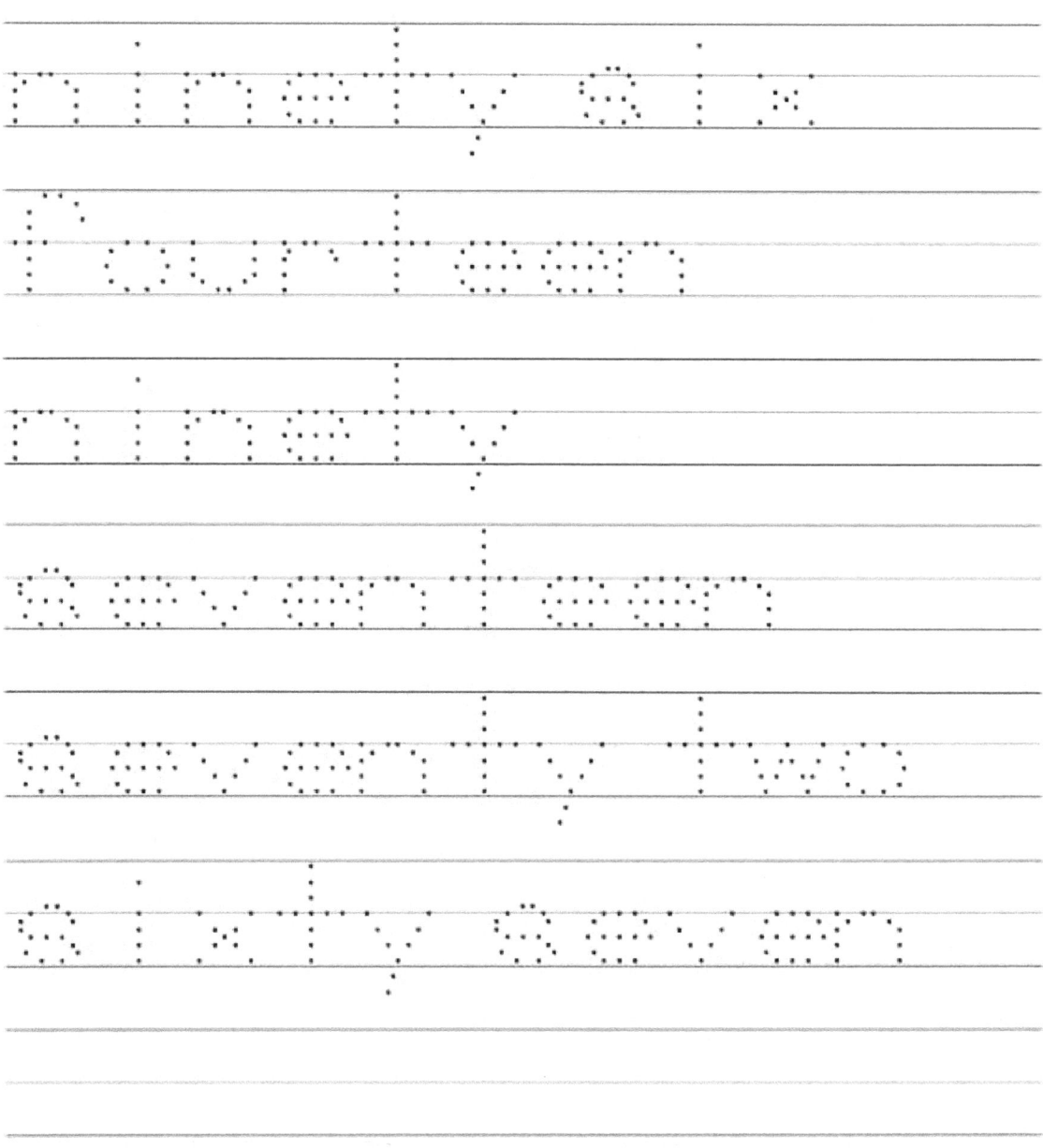

ninety six

fourteen

ninety

seventeen

seventy two

sixty seven

SMALL NUMBER WORDS EXERCISE
64

thirty four
sixty eight
three
seventy three
sixty nine
five

SMALL NUMBER WORDS EXERCISE
65

seventy five

twenty

thirty

five

seventy six

eleven

SMALL NUMBER WORDS EXERCISE
66

eighteen

twelve

forty nine

ninety two

twenty eight

seventy six

SMALL NUMBER WORDS EXERCISE
67

seventy seven

eighty three

eleven

thirty nine

seventy two

forty two

SMALL NUMBER WORDS EXERCISE 68

one

eighteen

eighty nine

sixty four

sixty nine

eighty six

SMALL NUMBER WORDS EXERCISE 69

forty one

seven

twenty three

sixty four

ninety three

fifty two

SMALL NUMBER WORDS EXERCISE 70

sixty nine

sixty nine

one hundred

five

sixty six

twenty four

SMALL NUMBER WORDS EXERCISE
71

three

twenty eight

eighty two

sixty three

nine

ninety two

SMALL NUMBER WORDS EXERCISE 72

twenty one

seven

eleven

seventy seven

one hundred

eighty

SMALL NUMBER WORDS EXERCISE 73

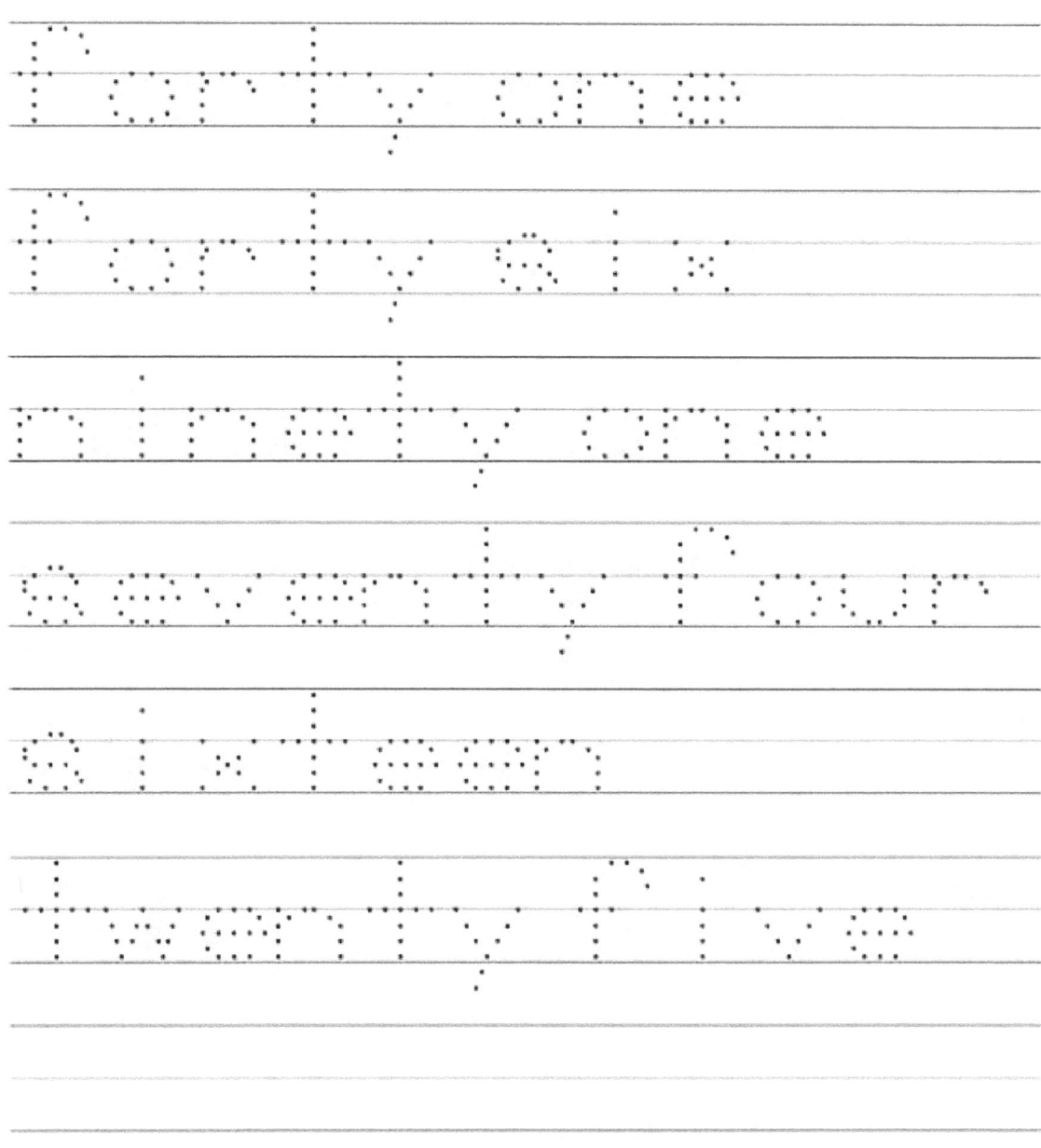

forty one

forty six

ninety one

seventy four

sixteen

twenty five

SMALL NUMBER WORDS EXERCISE
74

fifteen

fifty-two

sixty seven

eighty three

eighty eight

fifteen

SMALL NUMBER WORDS EXERCISE
75

twenty seven

sixty nine

ninety two

sixty

twenty nine

sixty five

SMALL NUMBER WORDS EXERCISE 76

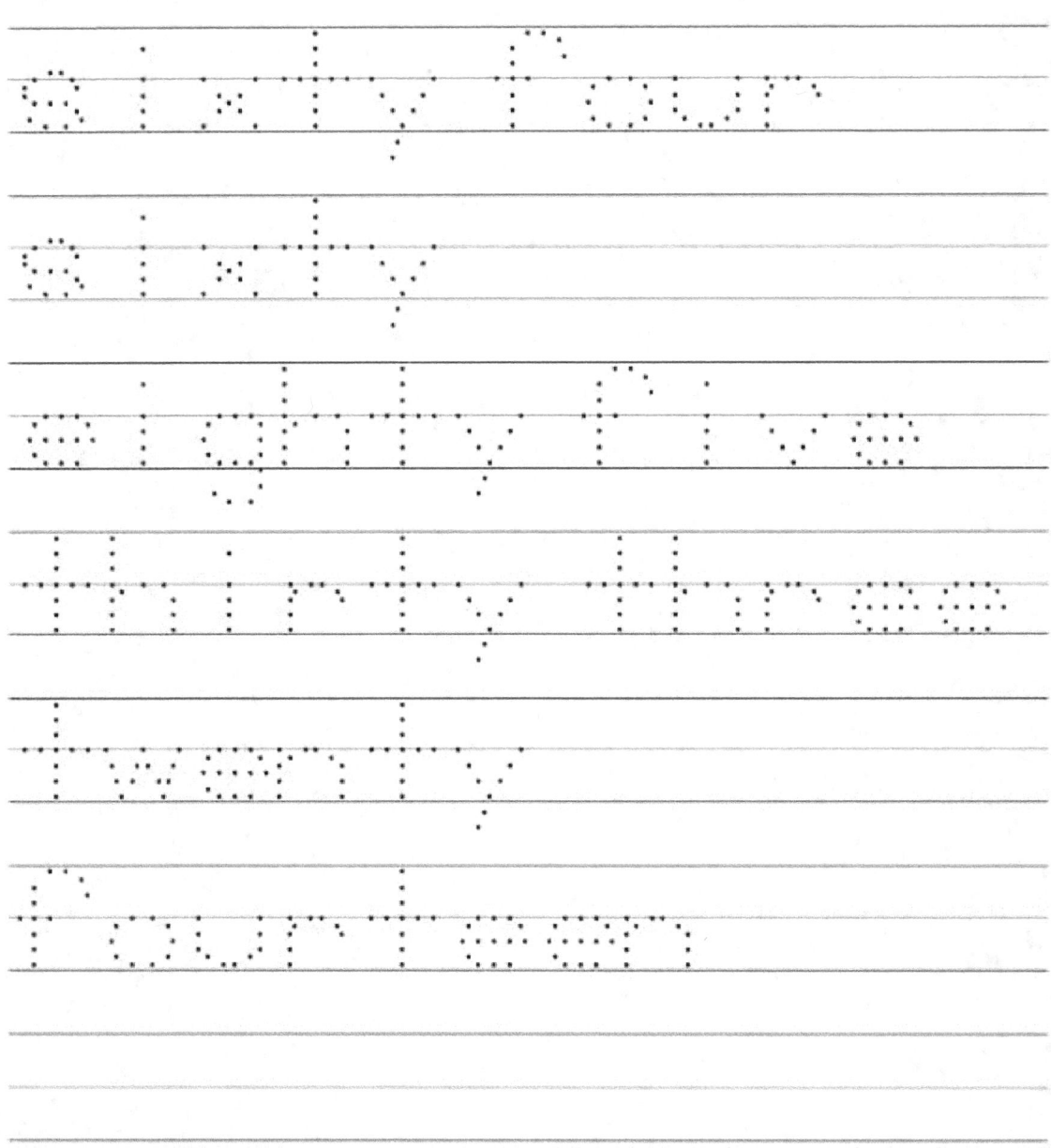

sixty four

sixty

eighty five

thirty three

twenty

fourteen

SMALL NUMBER WORDS EXERCISE 77

forty eight

seventy one

seventy nine

thirty two

sixty five

three

SMALL NUMBER WORDS EXERCISE 78

eighty three

seventy four

sixty four

twenty five

fifty four

sixty seven

SMALL NUMBER WORDS EXERCISE
77

forty eight

seventy one

seventy nine

thirty two

sixty five

three

SMALL NUMBER WORDS EXERCISE 78

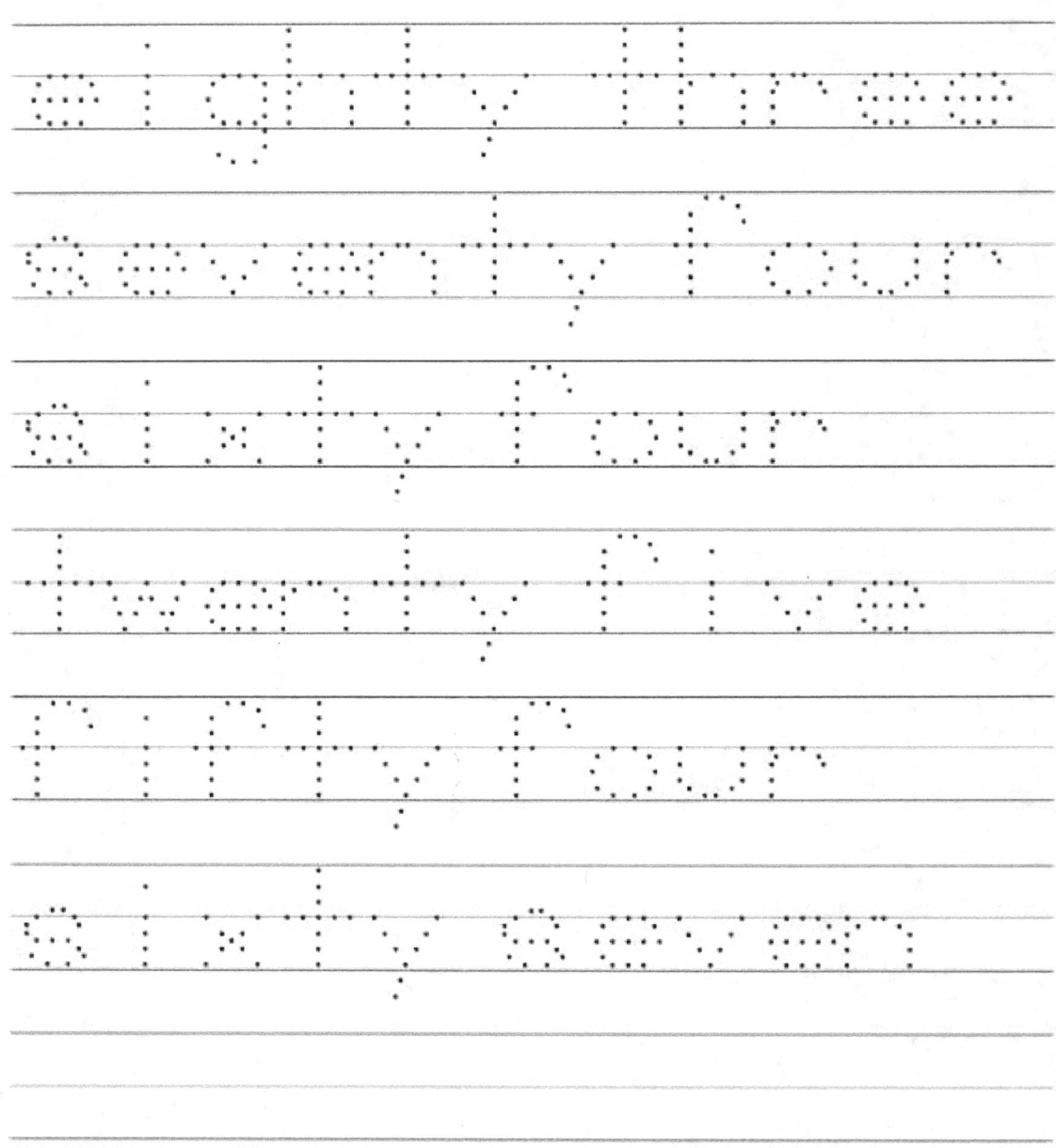

eighty three
seventy four
sixty four
twenty five
fifty four
sixty seven

SMALL NUMBER WORDS EXERCISE
80

seven

ninety two

twenty six

eighty three

fifty four

sixty eight

SMALL NUMBER WORDS EXERCISE 79

twelve

twenty nine

eighty eight

seventy six

six

forty one

TO SAY THANK YOU FOR PURCHASING THIS BOOK I AM GIVING AWAY A FREE COPY OF MY LATEST CHILDREN'S BOOK OR PUZZLE BOOK!

JUST GO TO THE LINK BELOWAND COOMENT WITH SITE LINK FOR BOOK REVIEW TO RECEIVE YOUR COPY

http://sudokuprintable.blogspot.com

TO SAY THANK YOU FOR PURCHASING THIS BOOK I AM GIVING AWAY A FREE COPY OF MY LATEST CHILDREN'S BOOK OR PUZZLE BOOK!

JUST GO TO THE LINK BELOWAND COOMENT WITH SITE LINK FOR BOOK REVIEW TO RECEIVE YOUR COPY

http://sudokuprintable.blogspot.com

CHECK OUT MORE BOOKS BELOW

All book collections can be found on

https://www.createspace.com/7511563

https://www.createspace.com/7512265

https://www.createspace.com/7512005

YOU MIGHT ALSO BE INTERESTED IN MY NEWEST COLLECTION OF BOOKS

CHECK OUT THE BOOKS ON THE NEXT PAGE

500 SUDOKU PUZZLES WITH ANSWERS

FIND OUT MORE HERE:

https://tinyurl.com/sudoku501

www.ingramcontent.com/pod-product-compliance
Lightning Source LLC
Chambersburg PA
CBHW082213220526
45470CB00010B/3149